The ABC's of Using Color in Rotationally Molded Products

Written by

Bruce Muller
President of
Plastics Consulting of South Florida
682 SW Falcon Street
Palm City, FL 34990

Copyright @ 2018

Bruce Muller

www.PlasticsConsulting.com

All Rights Reserved
First Edition

ISBN: 1986353273

Preface

Welcome to the Universe of Coloring Rotationally Molded Products

Selecting quality colorants and using those colorants properly will improve the Consistency, Quality and Durability of Rotationally Molded Parts. Rotomolded parts are colored for decorative purposes, identification and for functionality. Often additives are included in the color formula to improve properties of the rotomolded part.

Bruce Muller, President of Plastics Consulting, Inc. has written "The ABC's of Using Color in Rotationally Molded Products" for the rotational molding, compounding and colorant industries. The goal of this publication is to provide a basic overview of how to color products produced using the rotomolding process. A second goal is to educate rotational molding management that may not be familiar with all of the aspects of using color and the effects of pigments on molding cycles and the physical effects on finished parts. The intent of this book is to provide basic information on the methods of how to use colorants and the potential effects of using color. Much of the terminology unique to colorants designed for use in rotational molding will be discussed. Opinions expressed in this book are the opinions of the author.

Bruce Muller offers this information as a guide to help rotational molders and their suppliers expand their businesses, train their employees, educate their customers and develop beautiful, durable and innovative rotomolded products.

The ABC's of Using Color in Rotationally Molded Products

Chapter 1	**The History of using Color in Rotomolded Parts**
Chapter 2	**Forms of Colorants used to Color Rotomolded Resin**
Chapter 3	**Concentrate used to Produce Precolor**
Chapter 4	**Precolor or Compounded Resins**
Chapter 5	**Dry Color including Dustless Dry Color**
Chapter 6	**Paste Dispersions and Liquid Color**
Chapter 7	**Suggestions for Selecting a Color Supplier**
Chapter 8	**Dry Color and Liquid Color Usage**
Chapter 9	**Mixing techniques to mix Dry Color and Polish the Resin**
Chapter 10	**The Effects of Color on the Finished Part**
Chapter 11	**Requesting a New Color Match**
Chapter 12	**Trouble Shooting**
Chapter 13	**Glossary of Terms for Colorants**

Chapter 1

The History of Rotational Molding and the Coloring Rotational Molded Parts

The rotational molding process is very old, but the first recorded patent of a rotationally molded thermoplastic was issued to E.I. DuPont de Nemours and Company, J.H. Clewell and Rueben T. Fields in 1941. The patent describes molding plastisol, a liquid made by dispersing polyvinyl chloride resin (PVC) in liquid plasticizers.

In 1947 the B.F. Goodrich Company began to market a paste resin to produce plastisol. A Goodrich development engineer noted that *"Children's toys, including exquisitely molded dolls, can be made by a paste (plastisol) molding process."* We can assume that color was used to produce the doll heads and body parts. Paste Dispersion, the colorant used in the plastisol, was easy to develop for this low temperature liquid application. PVC items manufactured by other processes, had been colored using paste dispersions for several years.

In 1951, records exist of the first rotationally molded powdered polyethylene. We can assume that as soon as a rotationally molded polyethylene product reached the market place, quality color would have been a desirable attribute.

In the fifties and sixties several colorant manufactures existed, coloring injection molded, extruded, compression molded parts and calendared thermoplastic sheeting and fabrics. It has not been easy for these manufacturers to develop colorants for the newer rotomolding process with higher temperatures and longer cycle times.

Another problem encountered by the early colorant manufacturers, was to develop colorants for powdered polyethylene in a low shear (no shear) process versus for the pellets molded in the high shear processes they had historically provided color for.

A learning curve still exists today for many of the colorant manufacturers, as their labs don't have the proper equipment to rotomold parts and test equipment to determine how the colorants effect the physicals of the rotomolded parts.

These early colorant manufacturers also had to adjust to develop colorants that weathered well outdoors. Many rotomolded products, like tanks, playground equipment, stadium seating and marine products are used outdoors. The products they historically colored were molded for indoor applications like housewares, garbage bags and small injection molded toys.

Chapter 2

The Forms of Color used to Color Rotational Molding Resin

There are four forms of colorants used to color thermoplastics.

> 1. **Concentrate or Masterbatch is used to Manufacture Precolor or Compounded Color for Rotomolding. Below are examples of the Concentrate pellets.**

Strand Cut Masterbatch Pellets

Underwater Cut Masterbatch Pellets

2. Precolor or Compounded Color before Pulverizing for Rotomolding

Strand Cut Precolor Polyethylene Pellets

3. **Dry Color (fine powder) and Dustless Dry Color (granular)**

Dry Color is a Powder

4. **Paste Dispersions and Liquid Color**

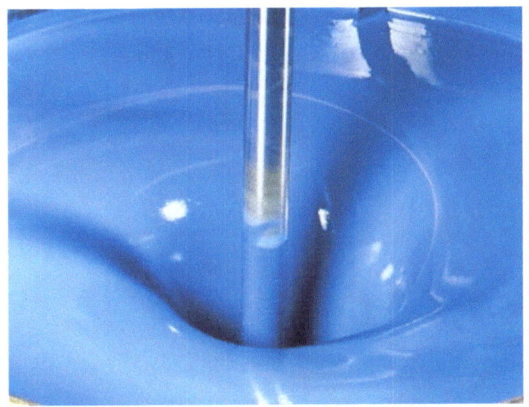

Liquid Color and Paste Dispersions

Each form of colorant has advantages and disadvantages. These will be discussed in detail in the following chapter.

Chapter 3

Masterbatch or Concentrate is used to Produce Precolor

Concentrate or Masterbatch is a concentrated form of colorants and additives, typically in pellet form. The pellets range in size from about ¼ x ¼ (0.250 x 0.250) inches down to 0.016 x 0.016 inches. The pigments, dyes and additives are concentrated with loadings, often as high as 65%, in a resin matrix. Masterbatch cannot be used directly in the rotational molding process, but are often used to produce compounded or precolored resin. The masterbatch is typically mixed at ratios of 2% (50-1) to 5% (20-1) with natural pellets and then extruded into a pelletized compound. The color compounded pellets are then ground into 35 mesh powder to process properly in the low shear rotomolding process.

Some advantages of masterbatch used to produce precolor are:

- The pigments and additives are well dispersed
- Higher pigment loadings are possible with less effect om physicals
- Pellets are dust free
- Pellets are easy to convey with air or vacuum
- Dry color mixing is not required in the rotomolding shop
- Most concentrate manufacturers produce quality pigment dispersions

Some disadvantages of masterbatch used in the manufacture of precolor are:

- The pigments, additives and resin matrix have a minimum of 2 heat histories prior to pulverizing. This potentially reduces the heat stability and weatherability of the finished product
- Masterbatch is more expensive than dry color or liquid color to produce the same Precolored compound
- Masterbatch lead times are much greater than for dry color or liquid color
- Masterbatch for polyethylene and polypropylene often contain high melt (20 - 30 melt) non stabilized LDPE that may have a small detrimental effect on higher density and lower melt index rotomolding grades of resin by potentially reducing some of the physicals and outdoor weatherability.

Chapter 4

Precolor or Compounded Resins

Precolor or Compounded resins are extruded into pellet form and are then pulverized into rotomolding grade powder. Precolored resins are typically compounded on either single or twin screw extruders.

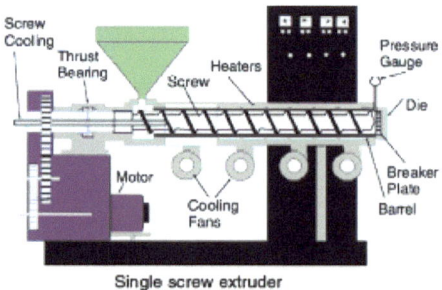

Single screw extruder

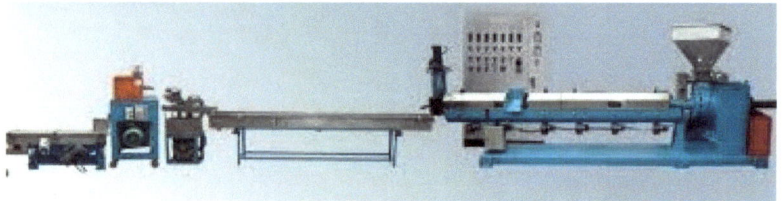

Extrusion Compounding Line

Precolored resins may contain pigments in polyethylene, polypropylene and dyes and pigments in Nylon and Polycarbonate. Additives such as antioxidants, UV absorbers, antistats and dispersants may be used in all rotomoldable polymers to promote additional beneficial properties such as heat and weather stability.

The advantages of Compounded Color for a rotomolder are:

- Potential for improved part opacity
- Precolor is the method used to produce granite colors. Larger pulverized particles of different of colors are used to produce the multicolor granite effect
- Precolor is quicker and easier to change colors than dry color or liquid color
- Pigment weighing scales are not required
- Color mixing equipment is not required
- Color consistency may be better than in house coloring due to larger lots
- Reduced opportunities for contamination due to less in plant resin handling

The disadvantages of Compounded color are:

- Higher cost
- Higher shipping cost
- Longest production lead times compared to liquid color and dry color
- Requires inventory space for each color – silos may be impractical for a molder using a multitude of colors or polymers
- Difficult to predict color obsolescence may create expensive obsolete inventory
- Pigments, additives and the resin have two or more degrading heat histories prior to rotomolding that may reduce heat stability and part life
- Polyethylene precolors are often made with masterbatches that contain non stabilized low density high melt, from 20 to 30 MI, non rotomolding grades of resin as the matrix (pigment carrier)

Chapter 5

Dry Color and Dustless Dry Color

Dry Color is a colorant in powder form. Dry color is an intense mix of pigments and additives, typically produced in a high intensity mixer.

High Intensity Mixer Used to Manufacture Dry Color

High Intensity Mixer Blades

Dry color can be used to produce precolor instead of concentrate, but is seldom used in that fashion. Dry color, used at reasonably low levels, is recommended to be high intensity mixed with natural powdered resin in the rotomolding plant.

The advantages of dry color:

- It is very low cost compared to compounded color or precolor
- Lead times for ordering dry color production may be very short for previously matched colors
- Usually can be shipped quickly using UPS or FedEx for low cost freight, and as little as 100 pounds of dry color may color up to 45,000 pounds of resin
- Requires small storage space and is used with natural resin
- Natural resin can be stored in silos, boxes and hoppers which can be conveyed to the mixers to automate part of the process
- Often one color may be used with different grades of resin
- Dry color is available from most manufacturers in pre weighed bags referred to as units. Units may be any size of pre-weighed color to eliminate weighing the dry color in the plant. The units are usually sized to color the resin in a batch determined by the mixer size

The disadvantages of dry color:

- It can be dusty and sticky requiring careful cleaning of the handling and mixing equipment
- Care must be taken to protect employees from inhaling pigment dust
- Cross contamination is always a possibility with airborne powdered color
- Dry color pigments and additives are not dispersed as well as compounded color or liquid color
- Accurate weighing equipment (a scale) is required

- High intensity mixing equipment is required
- Low levels of color are required to maintain part physicals, thereby restricting part opacity, especially in thin and single wall parts
- Over use of dry color may cause plate out in the mold
- Over use of dry color may allow free color on the inside of the part
- Adding some additives in dry color form, like UV absorbers, are less effective than adding them in compounded color. This is partly due to better dispersion and additive distribution in the compounded color
- Warping pigments may be more aggressive in dry color form than in compounded color

Chapter 6

Paste Dispersions and Liquid Color

Paste Dispersions and Liquid Color are of course in liquid form. Both liquids are the ultimate in pigment dispersion. These liquid dispersions are typically made on the same equipment used to manufacture paint, which disperses pigments and additives extremely well. The better the pigment dispersion, the brighter the color and the better the opacity. Optimum dispersion of pigments generally means lower color cost.

Paste Dispersions are used to color plastisol (liquid vinyl). Paste dispersions are as old as vinyl products themselves. Paste dispersions are simple to manufacture. Plastisols contain liquid plasticizers that are often used as the matrix to disperse the pigments in to produce PVC paste dispersions. The pigments and additives are easily dispersed in the plastisol compatible plasticizers, which do not affect the part physicals or processing and are generally low cost.

Liquid Color for polyethylene was very difficult to develop, as polyethylene does not contain liquid components like plastisol. Therefore, the liquid carriers used in liquid color for polyethylene, must be compatible with the resin, non-toxic, must not be flammable and must be a reasonable cost. It is advantageous for liquid colors to be cleanable with soapy water in preference to using hazardous solvents for cleanup.

A few advantages of liquid color:

- Liquid colors are low cost to manufacture
- Pigment and additive dispersions are excellent often reducing the color use cost
- Liquid color is easy to automate with peristaltic pumps
- It is dust free and the liquid carriers are generally nontoxic and non-flammable
- Weighing equipment used for dry color is generally suitable to weigh liquid color
- Mixing equipment used for dry color (high intensity mixers) is generally suitable to mix liquid color
- Inventory space is minimal since the very high strength liquid color is shipped in 5 gallon pails or 30 gallon drums
- Liquid color is very strong requiring only about 400 pounds shipped in a 30 gallon drum, to color up to 45,000 pounds of resin
- Often suppliers allow return of the empty drums to be refilled eliminating container disposal issues

A few disadvantages of liquid color:

- Liquid colorants should be inventoried near room temperature, because the viscosity changes up and down with variations of the ambient temperature. Very low viscosity may allow the pigments to settle out and very high viscosities make liquid color difficult to pump and meter
- If spilled, liquids can create a cleaning problem
- Purchase of metering equipment may be required to automate pumping and metering liquid colors
- If high intensity mixers are used, equipment cleaning may be more difficult

- Part testing should be performed to confirm the liquid matrix is compatible with the rotomolding grade of resin, not detrimentally affecting the physicals and outdoor life of the part
- There are only a few liquid colorant manufacturers in the US, therefore liquid color may be shipped from suppliers that are not close to the rotomolding plant

Chapter 7

Selecting a Colorant Supplier

Hundreds of colorant suppliers exist in North America. Unfortunately, only a handful have reasonable knowledge of the rotational molding process and have invested in a lab sized rotomolder.

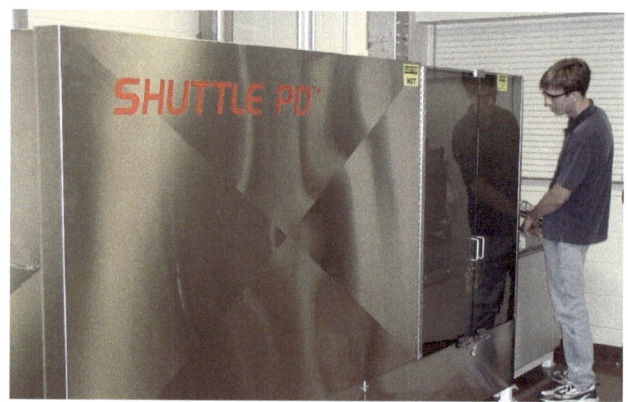

The Research Rotational Molder at
Pennsylvania College of Technology

Mold used to Make Test Parts
12" x 12" mold has Flow Gauges and a Shrink Gauge

Research Test Mold to Produce 5" x 5" Impact Plaques. The mold contains a Shrink Gauge and fill port for multi-wall moldings of different colors or polymers

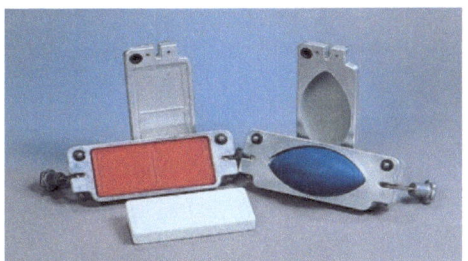

Molds Designed for Color matching and Promotional items

Purchasing from a colorant supplier without the proper rotational molding equipment and test equipment is very risky. Lack of knowledge of the process, by the colorant manufacturer, will increase the risk of the rotomolder manufacturing substandard parts due to the pigments and additives affecting part physicals. Poor color formulations may also adversely affect the oven processing cycles, cause warpage and early mold release.

Below is some of the test equipment your colorant manufacture should own and use in addition to impact test equipment that will be discussed in Chapter 8.

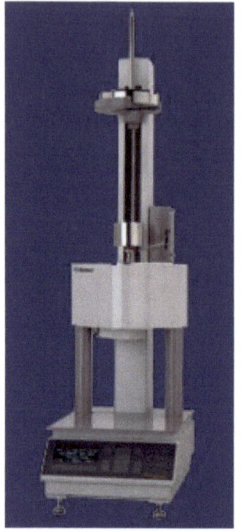
Melt Indexer

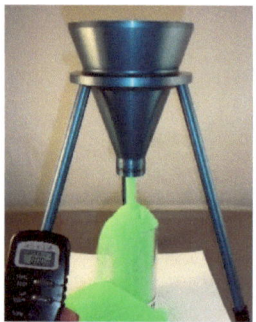
Flowability Funnel

Ro-Tap Sieve Analyzer use to determine particle size distribution of the powdered resin

Pigments and additives, affect cycle time, warpage, bleeding, outdoor weathering and the impact of rotomolded parts. Not only do the type of pigments and additives effect the parts, but also how well the colorants are mixed have an effect. The levels of colorants used will affect the oven cycle time and brittleness of the finished parts. This is especially true when using dry color.

Chapter 8

Dry Color and Liquid Color Usage

There is a very old industry guide line for the **maximum dry color** usage in rotational molded polyethylene. Other resins will require other usage guide lines. I call the polyethylene rotomolding guide line for usage "**the Old Rule**".

"Never use more than 100 grams of dry color in 100 pounds of resin."

This a very good rule in general, but each color must be evaluated for proper usage. Over usage of dry color in polyethylene will cause processing problems, scrap, color bleed, mold plate out and potential field failures. Often, 100 grams of color per 100 pounds of resin is excessive and will cause impact problems. Molders must depend on a knowledgeable colorant supplier to recommend **a not to exceed dry color usage level**.

There are many **exceptions that require lower usage**

- Colorants that contain only carbon black, usage levels may be as low as 25 or 30 grams per 100 pounds of resin depending on the type or grade of black
- Colorants that contain high levels of ultramarine blue, usage levels may be as low as or lower than 50 grams per 100 pounds of resin
- Colorants that contain high levels of organic pigments, like phthalo blue, phthalo green and many organic red, orange and yellow pigments, safe usage levels may be as low as 45 grams per 100 pounds of resin

There are several possible **exceptions for higher usage of Dry Color:**

- If a colorant contains fluorescent pigments a higher usage maybe OK
- If a colorant contains pearl or metallic pigments a higher usage maybe OK
- Tests have been performed on from 3% to 5% phosphorescent pigments dry blended into rotomolding resins with very little loss of impact
- If a colorant contains high levels of UV absorbers or antioxidants a higher usage may be OK
- Fluorescents, pearls, metallics and phosphorescents have very low oil absorbsion and don't require significant pigment wetting by the resin. Therefore, they may be used at levels exceeding **"the Old Rule"** of not exceeding 100 grams of dry color per 100 pounds of resin

Determining the Proper Dry Color Usage

The problem is that the rotomolder doesn't have a quick and easy method to determine which pigments and their levels are used in each colorant formulation. Therefore, a rotomolder must rely on their colorant supplier to recommend the proper maximum usage of each color. In order for the supplier to make a proper recommendation, they must maintain and use a laboratory rotomolder and lab test equipment to determine the proper colorant usage.

One of the best methods for a rotomolder and their colorant supplier to test for proper color levels is to test their parts with an ARM 5.2 Low Temperature Falling Dart Impact Tester. Polyethylene plaques are **conditioned** to -40° C for three hours.

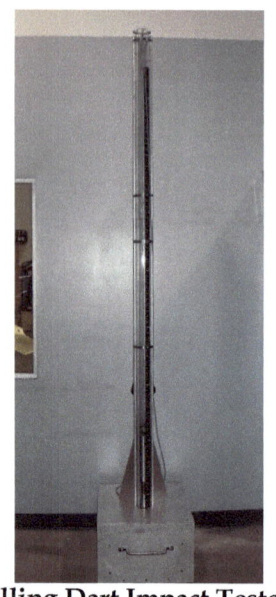
Falling Dart Impact Tester

Sample Holder in Tester

Weight Set

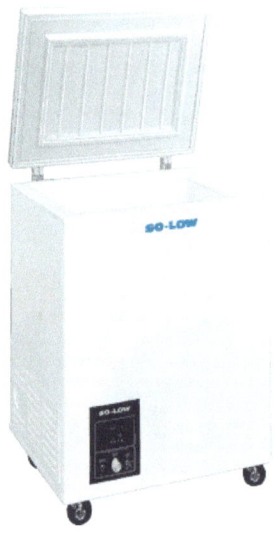
Freezer to condition Plaques to -40° C

The protocols for conditioning for the test are ARM 5.2 Low Temperature Impact and ASTM D3029 Method G. Excessive pigments will cause brittle parts. Under cure and over cure will also cause brittle parts, so care must be taken to determine if brittle test results are due to over pigmentation or under cure (under densification) or over cure.

The usage levels of pigments are affected by:

- The quality of the pigment and resin mix. The opacity, the pigment strength and brightness are affected by the intensity of the mix, the first step in proper dispersion
- The oil absorbsion of the pigments or the ability of the pigments to wet out by the resin, which is the second step in proper pigment dispersion
- The method used to manufacture the pigments – pigment oil absorbsion of similar pigments will vary from different pigment manufacturers
- The density of the resin – the lower the resin density the better the ability to wet the pigments. Generally, density of 0.935 will wet the pigment better than 0.95
- The melt index of the resin – the higher the melt index (the lower the molecular weight distribution) the better the wetting ability of the resin
- The surfactant level in the colorant. Unfortunately, the best pigment surfactants may cause warpage and are used at very low levels in rotomolded polyethylene

Chapter 9

Mixing Techniques to Mix Dry Color and Polish Resin

High intensity mixers and high speed paddle plow mixers are the most widely used equipment to obtain the best mix of dry color into powdered polyethylene.

High Intensity Mixer 150 liter mixes 100 lbs. of 35 mesh PE

High Intensity Mixer 150 Liter

Paddle/Plow Mixer – 800 Pound Capacity

Internal views of both mixers

High Intensity mixer blades and thermocouple top left

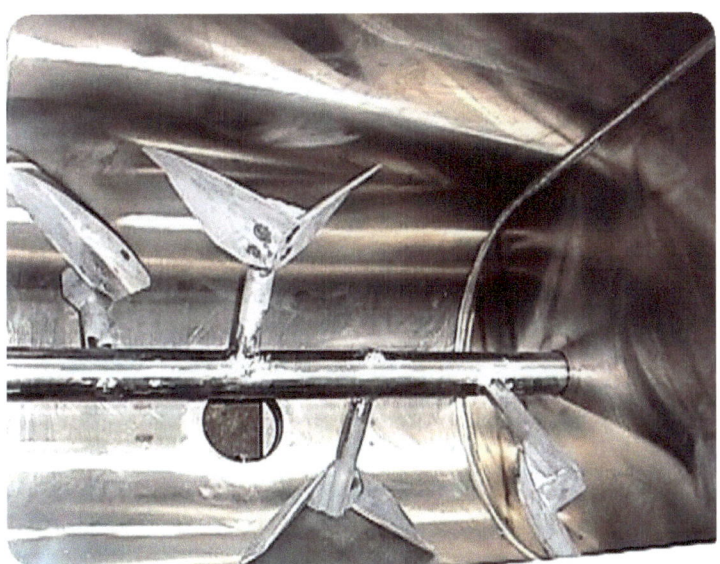

Stainless Steel Paddle Plow mixer blades and discharge port

A high intensity mixer creates shear or friction. It is that shear that accomplishes the proper mix of the colorant and the powdered resin. The shear creates frictional heat. It is always preferable to mix powdered resin and dry color by temperature rather than time. The temperature rise in the high intensity mix is the indicator of the amount of shear developed during the mixing cycle.

A high intensity mixer only has one proper resin fill level. If the mixer is under filled with resin, the shear will be reduced due to fluidation, and the mix may not be satisfactory or complete. If the mixer is over filled, the batch may not turn over in the mixer properly yielding an incomplete mix. My suggestion is to mix to 130° F (55° C). This should take from 3 to 5 minutes in a good high intensity mixer and about 20 minutes in a large paddle/plow mixer. An added benefit to mixing to 130° F is that the resin will become polished. Polished resin may reduce cycle times as it will densify more quickly.

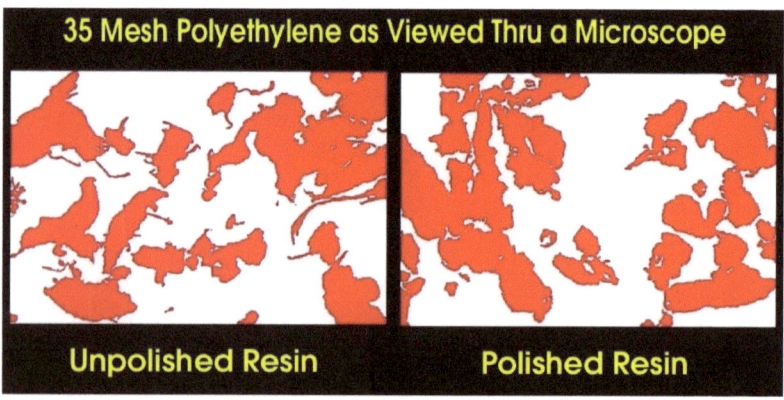

Polished resin will flow better in the mold and provide other benefits:

- Reducing the oven time. Tests have shown oven times reduced up to 14% due to improved sintering, due to less trapped air in the powder from rotating in the mold prior to laying up on the hot mold
- Fills around inserts and molded in threads better
- Fills in fine mold details
- May reduce airborne dust
- Potentially improve part impact due to better and quicker densification
- Allow easier filling of small molds as polished resin has a higher bulk density

There are no disadvantages to polishing resin as it is accomplished as the dry color is mixed properly into the resin, by frictional heat or temperature. Polishing to temperatures as high as 160° F (70° C) have shown no adverse effects on the resin.

Precolor, compounded and natural resins may be polished as an additional step after pulverizing when improved powder flow in the mold is desirable. Improving the flow in the mold by polishing the resin must justify the additional cost of the additional processing (polishing) step.

Chapter 10

The Effects of Color on the Finished Part

It is important to use color when a mold is first evaluated because colors effect the part quality, part size and the molding cycle. Different pigments effect how quickly the resin densifies, therefore affecting the cycle time. Studies have shown that white pigments (TiO^2) or colorants containing high levels of TiO^2 may require up to a 14% longer oven cycle than most other pigments. Unfortunately, many custom molders must mix colors on an arm to maintain optimum productivity. Even so, it is prudent to keep this information in mind when scheduling production, mixing colors on an arm and testing new molds.

Colorants containing pigments and additives with different chemistries may effect:

- Cycle time - densification
- Shrinkage
- Warpage
- Resin crystallinity
- Mold release ability
- Mold plate out
- Pigment bleed
- Reduced outdoor life
- Increased outdoor life
- Part cost
- Part impact
- Part toxicity
- Bubbles in the parting line
- Presence of static swirling

Chapter 11

Requesting a New Color Match from your Supplier

It is important to supply complete information to your color supplier when requesting a new color match. Not only is the information important, but they also must have a sample of your resin for the job to work up the match. Check with your color supplier for the quantity they require, but it will probably be in the range of 10 to 12 pounds per color match, depending on how large their test mold is and how many trials are needed to achieve a quality match. Some manufacturers return only a chip for approval while others automatically supply a small sample for a trial in your plant. Of course, the sample is ideal, but a sample for 100 pounds or so may create a problem to achieve a proper sample color mix with the resin, if you have only a large mixer. This may create a color difference compared to a larger production mix.

A flat plastic sample is always preferable for a colorant supplier to match to. It should be at least 3" x 3" square, large enough for a spectrophotometer to read. It is very difficult to achieve quality color matches with things like yarn, buttons and scraps of paper. Pantone matches are also difficult for color suppliers, because Pantone colors are matched with printing inks that cannot be used in high temperature rotomolding.

This is a list of information the color supplier will require to produce a quality color match:

- Address to send the chip and sample to
- Date required - **ASAP is not an appropriate date**
- Color name and job number. These are stored with their formulas and may be difficult to change later
- Sample size and number of color chips required
- Form of color required, for example dry color, compounded or liquid color
- Resin that will be used. Manufacturer, density and melt index are required
- Part thickness (thinnest area) and if it will be single or double wall as this may affect the opacity required
- Out door or indoor use
- If the standard needs to be returned
- Toxicity requirements i.e. heavy metals allowed or food application.
- The light source you will approve the color in. For example, day light, incandescent or fluorescent light

Chapter 12

Troubleshooting Guidelines

Static Swirling is created by static build up in the resin. It affects the smaller resin particles more than the larger resin particles. The resin particles build up a static charge during mixing, conveying and movement in the mold, especially in low humidity conditions. The smaller dry colored resin particles, that stick to the mold first, have greater surface area and hold more color than the larger particles and therefore appear darker. The larger particles move to the inside of the part and do not show swirling. Static swirling is also possible in precolored granite colors.

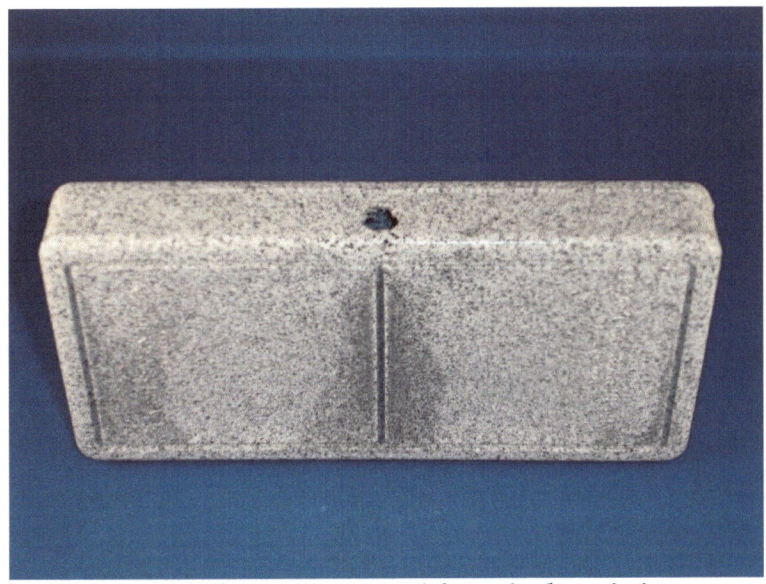

Rotomolded granite part with typical static issues

Static build up occurs during powder transfer, during mixing and during powder rotation in the mold. Swirling is easy to identify as it is usually in the same location on the part. It is often after a ridge in the mold and near the part corners as in the example above.

Static in rotational molding powdered resin may be reduced by:

- Slowing the mold rotation, but not changing the ratio
- Adding an In Process Antistat to the mix
- Allowing resin from the mixer to rest allowing static dissipation before use
- Ground the mixer with a high surface area wire – a copper water pipe is an excellent ground to attach the ground wire to
- Confirm there is good contact between the spider and the arm
- Confirm there is good contact between the mold and the spider
- Mold to spider grounding devices are available from some mold manufacturers

Brittle parts and parts with poor impact may be due to excessive pigment, especially when using dry color. Reduce the colorant in 10% increments until the impact failure becomes ductile. Remember that brittle parts may also be caused by over cure and especially under cure (under densification). Any bubbles in the part wall are an indication of poor densification (under cure).

Falling Dart Impact Brittle Failure

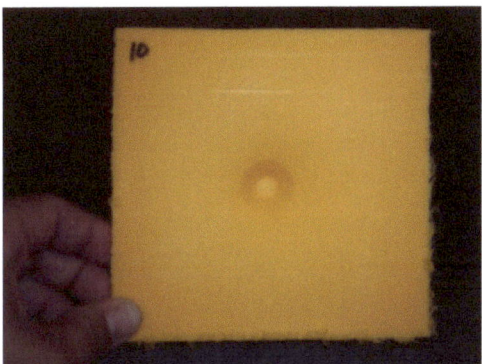
Falling Dart Impact Ductile Pass – Dimple holds water

Falling Dart Impact Ductile Failure - will not hold water

Marking the mold side of the impact plaque, using a Sharpie before conditioning, with the required information will avoid confusion recording results. When performing the ARM 5.2 Low Temperature Falling Dart test, it is important to precondition the 5" x 5" (12.7 mm x 12.7 mm) plaques for 3 hours @ -40° C (-40° F). A CD rack is perfect to hold the impact plaques vertical in the freezer. The plaques must be impacted on the mold side. I recommend impacting them within 8 seconds after removal from the freezer, as the plaque temperature increases quickly. Remove the plaques one at a time and close the freezer lid in between plaque removal. Marking the plaques on the mold side results and helps plaque orientation for impacting (mold side up).

Warping of Parts may be caused by certain organic pigments, waxes and surfactants like Zinc Stearate. Phthalocyanine blue and green are well known warping pigments in polyethylene. There are several other organic pigments that are prone to cause warpage also. If you plan on molding parts that you expect to be prone to warping, notify your colorant supplier so they can avoid using known warping pigments in your colorant. Or, if you are experiencing warping problems, notify your color supplier for a formulation change. Warping may also be caused by cooling the mold too quickly. I recommend misting the molds with water after a few minutes' delay. As long as the part is in contact with the mold it will remain in the shape of the mold. As the part releases from the mold it will shrink and change shape. Unfortunately, the part does not release from the mold all at once and therefore does not cool uniformly. Part cooling is reduced drastically after the part releases from the mold. The warpage in the picture below is caused by an improper part design and possibly phthalo green pigment in the colorant.

This warped part is due to poor design and Phthalo green pigment (the warping pigment)

Mold release that is overly aggressive will also cause warpage. A permanent mold release like Teflon®, when too aggressive, can be roughed up with a nylon scrub pad to reduce the release properties. The Teflon coating can be roughed up on the entire surface or only in a troublesome area. This is a permanent change, so care must be taken.

A semi-permanent release may be changed for a less aggressive type of release or stripped and reapplied more uniformly. Most water based mold releases require a bake cycle to thermoset them. Thermosetting the mold release will increase its life cycle.

Pigment Bleeding can be a serious problem. Clothing, carpets and even skin may be affected by inferior pigments bleeding out of rotomolded parts. Many pigments that are very permanent in high shear polyethylene processes like injection molding and extrusion may have a tendency to bleed out of rotomolded parts.

A test method used to identify bleeding pigments, used throughout the rotational molding industry, is attached. During the development of this test, it was learned that the bleeding or lack of bleeding, found after 24 hours, did not significantly change for the next 10 months.

It is recommended to test only room temperature parts that have thermally stabilized.

TECHNICAL BULLETIN B-17

BLEED TEST FOR POLYETHYLENE COLORANTS used in Rotational Molding

The Problem
Partly due to the extreme low shear during the rotational molding process, many pigments that are normally permanent in polyethylene, when molded by other processes, will bleed from rotomolded parts. The bleeding is fairly immediate after molding and will continue for the life of the part.

The Risk
Outdoor playground equipment like slides can stain not only children's clothing, but also their skin. Rotomolded toys used indoors have stained carpet and the toy manufacturers were sued for carpet replacement. Bleeding color has discolored and contaminated fluids stored in rotomolded tanks.

The Bleed Test
This test procedure will identify and differentiate bleeding colorants from poorly dispersed (mixed), floating or poorly wet out colorants[**].

The Bleed Test Procedure
Spray an area about 5" x 5" on the **inside surface** of a single wall part with **Fantastic®** household cleaner[***]. Rub the sprayed area briskly with a white facial tissue (Kleenex®). Even a non bleeding colorant may show slight traces of color on the tissue. Spray the **outside** (mold side) of the part

with **Fantastic®** household cleaner***. Rub the sprayed area briskly with a clean white facial tissue (Kleenex®). There should be **no signs of color** on the tissue after rubbing. Studies have shown that results of the bleed test, performed 24 hours after molding, will not change significantly with time.

Evaluating the Results

High levels of pigment wiped off the inside of a part may indicate insufficient colorant mixing and/or colorant usage to high. Remember the **"Old Rule"** of not exceeding 100 grams of dry color per 100 pounds (45 Kilos) of powdered polyethylene. There are many exceptions to this rule, but it is a good guideline. Some of the colorants that may exceed these loadings are fluorescents, metallics, pearls, and phosphorescence pigments. Examples of pigments that must be used at **much lower levels** are phthalo blue and phthalo green, ultramarine blue, carbon black and most organic pigments.

Simplification of the Results

Bleeder colorant - color staining the tissue from rubbing the inside **and** the outside of the part.

Non Bleeder colorant – slight color discoloration of the tissue only from the inside part wall, with no discoloration of the tissue from the outside part wall. Excessive colorant loading or dispersion problems may also cause excessive color rub off, even when using compounded color.

*** Fantastic® has been proven to be a strong soap and works well for this bleed test
** It is not unusual to have some pigment rub off on the interior wall of a rotationally molded part

Poor Dry Color Dispersion, causing quality issues, from a colorant supplier, may be identified using a simple **Dry Color Smear Test.**

TECHNICAL BULLETIN B-16

DRY COLOR SMEAR TEST
Dry Color Draw Down Test

The Problem

Many Dry Color manufactures do not have the technology or the equipment to properly disperse dry color pigments for use in rotational molding. Proper dispersion is extremely important, due to the fact that mixing powdered resin with powdered dry color by a rotomolder, even in a high intensity mixer, develops very low shear compared to the intensive mixing and shear developed in the mixers used by Hi-tech Hi-quality dry color manufacturers.

Significance

Using dry color that has **not** been dispersed properly will cause:
- Low opacity
- Loss of color intensity
- Color variations (off color)
- Specking
- Low part impact
- Color rub off
- Bleeding and raw pigment on the inside of the part

The Dispersion Test

A very simple test exists to check dry color for proper dispersion. This test procedure is a method to determine if the dry color is free from poorly dispersed pigments and additives. Poorly dispersed pigments will show up as color streaks and even grit in the dry smear test.

The Test Procedure

Place a piece of coarse brown paper* on a smooth table top. ** Using a color chip*** or a 2 or 2 1/2 inch wide putty knife, holding the paper with your other hand, draw about one teaspoon of dry color down the brown paper. Use a lot of pressure and angle the putty knife in a way that the dry color is drawn down very thin for about 10 inches.

Observing the Results

Look for color streaks, caused by individual pigments that are improperly dispersed. Glossy shiny streaks are normal. With poorly dispersed pigments you will see streaks and hear grit during the draw down. The gritty pigments will appear as pencil streaks 3/8" long or longer in the draw down (smear).

Ready to begin

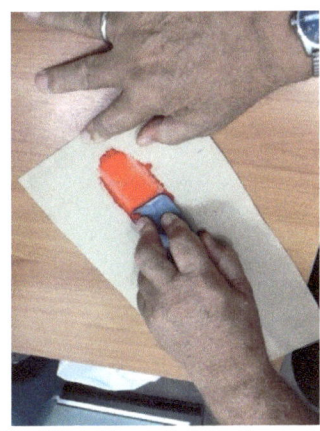
Starting the draw down

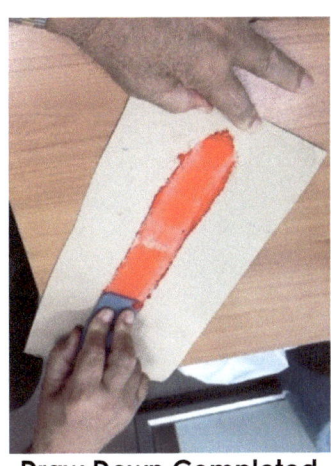
Draw Down Completed

Close up of a good draw down

A failed dry smear – note the orange, white and black streaks and specks

Tips for the Smear Test

If your dry color doesn't pass this test, **don't use it**. Using poorly dispersed dry color will cause color inconsistencies and color bleed. If your color supplier doesn't routinely pass this Dry Smear Test, I suggest finding another supplier. Once a quality supplier is found, this test only needs to be performed on the first lot of every new color match and when color problems occur in production.

Performing the **Dry Color Smear Test, is the first step to insure proper color, color consistency, part impact and scrap reduction**.

* A brown paper grocery sack is ideal. Tear the sack into single sheets. Brown paper can also be purchased on a roll. A piece about 10" X 14" long is suitable for about 4 dry smear tests.
** A Formica counter top or a piece of glass is a perfect smooth surface to perform dry smears tests.
*** A 1 ½" or 2" thin putty knife or a stiff injection molded polypropylene or styrene color chip supplied by your colorant supplier works well to draw down the Dry Color.

Chapter 13

Glossary of Terms for Colorants

Colorant
Any form of color used to color thermoplastics. The four forms of plastic colorants are Dry Color, Liquid Color (including Paste Dispersions), Concentrate or Masterbatch and Precolor or Compounded Color.

Colorimeter
An instrument used to evaluate color matches numerically. I consider these instruments to not be worth the investment. They will not identify Metamerism and therefore may allow mis-matched colors to be approved.

Color Wheel
The color wheel is useful in determining color progression and complimentary colors. The moniker for the color order is ROY G BIV or **R**ed, **O**range, **Y**ellow, **G**reen, **B**lue, **I**ndigo, **V**iolet.

Concentrate or Masterbatch

The terms are interchangeable. A form of color not suitable for rotational molding except to manufacture precolor or compounded color. It is normally in pellet form and is a dispersion of pigments and additives in a resin compatible with the resin it will be used in. The pigment and additive concentrations (loadings) vary from about 30% to 65% with the balance of the formula the matrix resin. Usage may vary from 1% to 5%.

Compounded Color

The same as precolor. It is normally resin mixed with concentrate (masterbatch) and then extruded into pellets. For rotomolding it is usually pulverized into 35 mesh powder. A very simple white compounded color formulation is below:

Rotomolding grade LLDPE	99.7 pounds
TiO^2 White pigment	85.0 grams
Ultramarine Blue Red Shade	2.6 grams
Surfactant	20.0 grams
	100.0 pounds

Dry Color

A powder form of colorant that generally contains no polymers, but may contain additives like UV absorbers, antioxidants, antistats and surfactants. A very simple white dry color formulation for 100 pounds of 35 mesh rotomolding LMDPE resin, approximately the same color as the compounded formula above is:

TiO² White pigment	85.0 grams
Ultramarine Blue Red Shade	2.8 grams
ZnSt Surfactant	10.0 grams
In process Antistat	<u>25.0 grams</u>
	122.8 grams

In this example the correct usage is 122 grams per 100 pounds PE. Note that the pigments and Zn St meet the **"Old Rule"** at 97.8 grams. The Antistat does not require wetting by the resin and therefore is exempt from the **"Old Rule"**.

Dyes

Are organic and always produced in powder form. Dyes dissolve in plastics and are generally very bright colors with very little opacity. Dyes cannot be used in polyethylene or polypropylene because they will bleed. Dyes are suitable for rotomolded nylon, polycarbonate, SAN and a few other polymers. A common use for dyes, that everyone is familiar with, are the red and amber colors used in automotive acrylic tail lights.

Light Booth

Often referred to as the Macbeth Light Booth. This is often a table top enclosure, with multiple light sources and a neutral colored surface inside, is used to evaluate colors, not only in plastics, but also fabrics, paint and printed materials. By visually comparing two colored objects under multiple light sources, a determination can be made as to how well a color will match under those different light sources. When a color matches under one light source and does not match under another, it is considered metameric.

A light booth may contain up to 5 different light sources including daylight (similar to June 23 at 12 o'clock noon), horizon light (reddish light like when the sun is setting), incandescent light (ordinary yellow light like a common cool white light bulb) and black light that will identify fluorescent pigments.

GretagMacbeth Judge II Table Top Light Booth

Liquid Color

Pigments that are highly concentrated in liquid form. Liquid color has a typical viscosity of 2,000 to 3,000 centipoise similar to 40 or 50 weight motor oil. Liquid color has the best pigment dispersion of the four forms of color, due to that fact many pigments, especially organics, develop into very high strength when processed with high shear. Liquid color is not commonly used in rotational molding, but certainly has been used successfully by many rotomolders. An efficient method to handle liquid color is to meter and pump it using a peristaltic metering pump. The pump transfers and meters the liquid color from the container into the mixer. High intensity mixers are suitable to mix liquid color into powdered resin. Before using liquid color in production, I suggest checking part physicals to make sure the liquid carrier is compatible with the grade of resin (polyethylene) you are using.

Metamerism or Metameric
When two colors match under one light source, but do not match under another light source, they are considered metameric.

Paste Dispersion
Paste dispersions have the same characteristics as liquid color. Often they are a little thicker or higher viscosity than liquid color. They are used in flexible vinyl or polyvinyl chloride (PVC), as they are manufactured with plasticizers already in the plastisols and flexible vinyl.

Pigments
Pigments may be organic, inorganic or specialty colorants. Pigments are in powder form and do not dissolve, but disperse in polymers. When the resin melts during the process, the pigments are wet out causing them to become bound in the resin after it cools.

Some examples of organic pigments are carbon blacks, phthalocyanines, and quinacridones.

A few examples of inorganic pigments are iron oxides,

titanium dioxide, cadmiums, ultramarine blue and mixed metal oxides.

Examples of specialty colorants are pearlescents, fluorescents, metallics and phosphorescents.

Precolor

The same as compounded color. Precolor is used in the rotational molding process in the same form as it is delivered. A few US rotomolders do extrusion compounding and many molders pulverize natural and compounded (precolor) resin in house. When a rotomolders resin consumption reaches higher levels it may become advantageous to pulverize in house.

Shrinkage

Polyethylene shrinks about 0.021 inch per inch. This will vary with resin density, part design, cooling rate, restrictions in the mold and many other factors.

A shrink gauge molded in a 12 x 12 inch test mold with ¼ x ¼ inch gauges

Spectrophotometer

An instrument used to determine color differences, formulations and pigment ratios numerically in a colored part. Spectrophotometers often are used in conjunction with computers allowing visualization of color curves and to develop formulations for a color match trial. They typically have the ability to read a color in three or more simulated light sources. Those light sources typically include Daylight, Incandescent and Horizon light. This ability allows for identification of colors that are non- metameric.

Outdoor light is not always daylight. As the sun sets and approaches the horizon or in the winter, the light becomes redder similar to the Horizon light simulated by the spectrophotometer. Incandescent light is similar to normal office and house hold cool white lighting.

Lab Scan XE by Hunter Lab

Hand Held Spectroeye by GretagMacbeth

References

Muller, Bruce, Plastics Consulting, Inc. TECHNICAL BULLETIN B-22, June 2013
"ASTM and ARM Testing and Materials Standards"

Muller, Bruce, Plastics Consulting, Inc. TECHNICAL BULLETIN B-17, June 2008
"A Bleed Test for Polyethylene Colorants".

Muller, Bruce, Plastics Consulting, Inc. TECHNICAL BULLETIN B-16, June 2010
"Dry Smear Test for Dry Color"

Beall, Glenn, Glenn Beall Plastics,
Photo of the green warped part on page 39

Pennsylvania College of Technology, Williamsport, PA
Photos of the SHUTTLE rotational molder and test mold on page 20

Contributors

Author: **Bruce Muller, Plastics Consulting of South Florida Inc., Palm City, FL.**

Cover designed by: **Bru Muller, Undercurrent, Los Angeles, CA.**

The information contained in this book is provided in good faith. Neither Bruce Muller nor it's contributors accepts responsibility for the application of this information. MSDS information must be provided by all material suppliers. Safe manufacturing procedures and practices must be followed at all times.

www.ingramcontent.com/pod-product-compliance
Lightning Source LLC
Chambersburg PA
CBHW040241220526
45473CB00001B/321